AF228520

SUPERHERO ANIMALS
BLACK PANTHERS
KENNY ABDO
Fly!
An Imprint of Abdo Zoom
abdobooks.com

abdobooks.com

Published by Abdo Zoom, a division of ABDO, P.O. Box 398166, Minneapolis, Minnesota 55439. Copyright © 2020 by Abdo Consulting Group, Inc. International copyrights reserved in all countries. No part of this book may be reproduced in any form without written permission from the publisher. Fly!™ is a trademark and logo of Abdo Zoom.

Printed in the United States of America, North Mankato, Minnesota.
102019
012020

Photo Credits: Alamy, Everette Collection, Getty Images, iStock, Shutterstock, ©Copeinator123 Fantastic Four Vol 1 52 p9 / CC-BY-SA
Production Contributors: Kenny Abdo, Jennie Forsberg, Grace Hansen
Design Contributors: Dorothy Toth, Neil Klinepier

Library of Congress Control Number: 2019941308

Publisher's Cataloging-in-Publication Data

Names: Abdo, Kenny, author.
Title: Black panthers / by Kenny Abdo
Description: Minneapolis, Minnesota : Abdo Zoom, 2020 | Series: Superhero animals | Includes online resources and index.
Identifiers: ISBN 9781532129490 (lib. bdg.) | ISBN 9781098220471 (ebook) | ISBN 9781098220969 (Read-to-Me ebook)
Subjects: LCSH: Black panther--Juvenile literature. | Leopards--Juvenile literature. | Big cats--Juvenile literature. | Phantom cats--Juvenile literature. | Predatory animals--Juvenile literature. | Carnivores--Juvenile literature. | Zoology--Juvenile literature.
Classification: DDC 599.75--dc23

TABLE OF CONTENTS

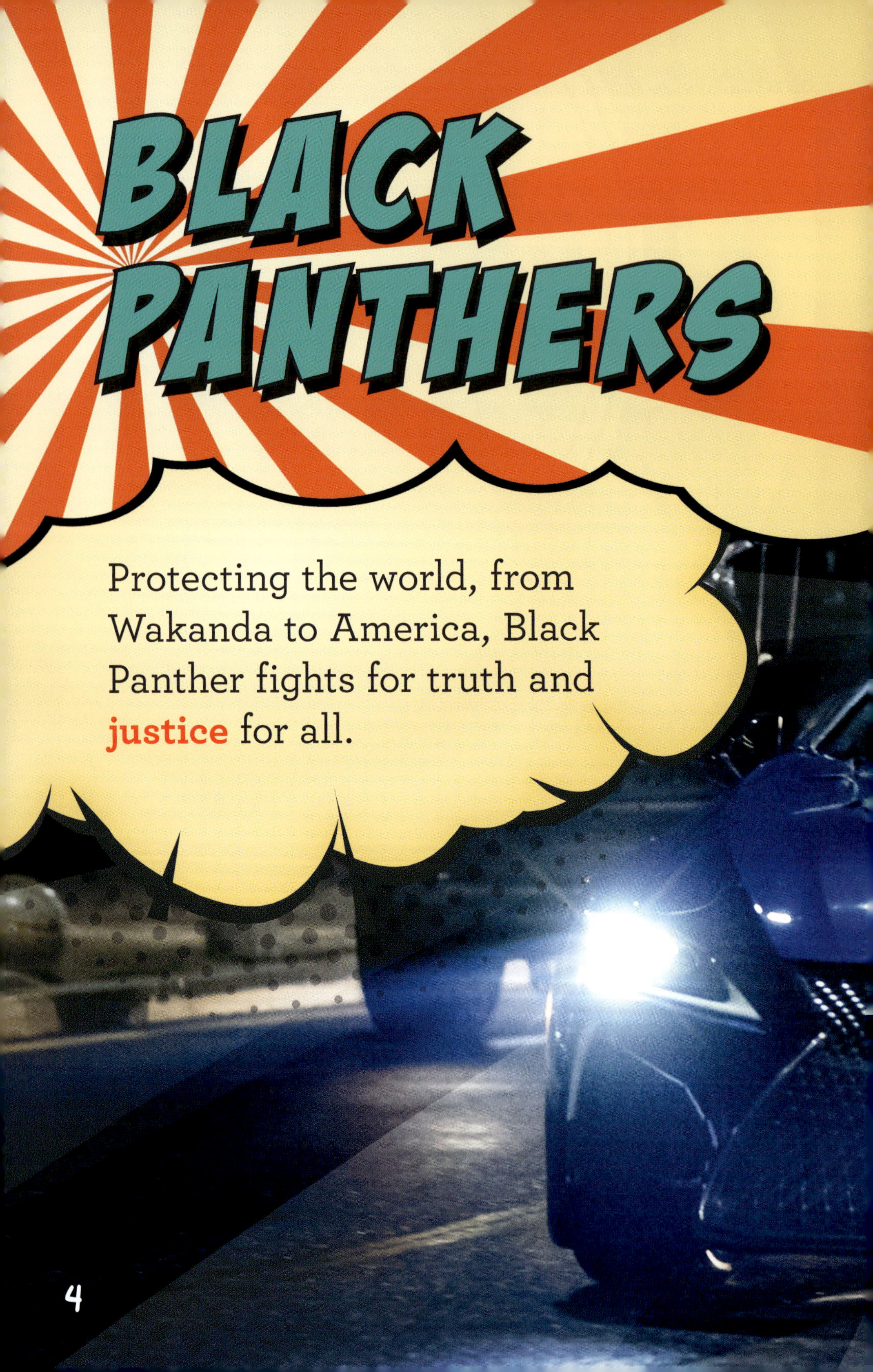

BLACK PANTHERS

Protecting the world, from Wakanda to America, Black Panther fights for truth and justice for all.

As one of the largest members of the big cat family, black panthers can be found roaming the jungles of Africa, Asia, and the Americas.

ORIGIN STORY

Black Panther clawed into action in *Fantastic Four* issue 52. Co-creators Stan Lee and Jack Kirby drew inspiration from many places.

Royalty throughout history along with famous **activists** brought King T'Challa to life. The fierce strength and intelligence of the black panther informed his powers.

The black panther is known as "the ghost of the forest." It is a smart, **stealthy predator**.

At night, its dark fur allows it to **camouflage** and stalk **prey** very easily.

Black panthers can run up to 35 mph (56.3 kph). This helps them catch fast **prey**, like antelope and deer.

Black panthers are strong swimmers. They spend a lot of time swimming, playing, and hunting in water.

They can quickly climb very high. Black panthers can jump from trees to catch their **prey**. They can leap about 20 feet (6 m) to nab monkeys, deer, and birds.

IN ACTION

Black Panther shares similar qualities to the animal. His suit matches the color of the **predator's** fur. His **senses**, like hearing and seeing, are heightened to a superhuman degree.

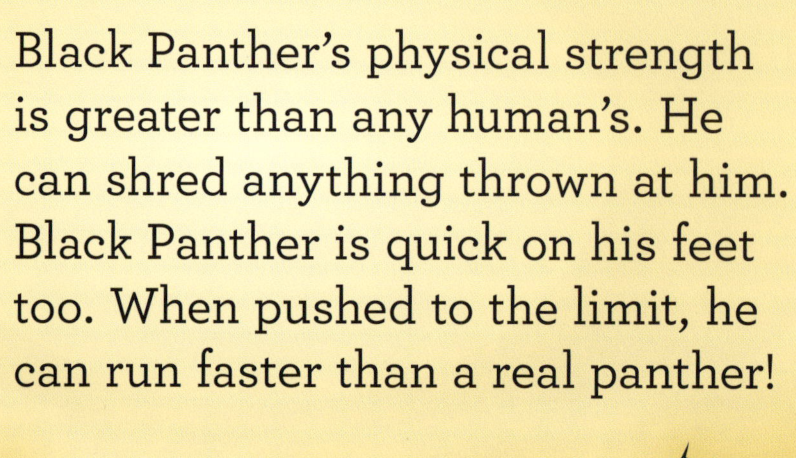

Black Panther's physical strength is greater than any human's. He can shred anything thrown at him. Black Panther is quick on his feet too. When pushed to the limit, he can run faster than a real panther!

T'Challa has a genius intelligence. He is **stealthy** and thoughtful. Like the black panther of the wild, he is a strong warrior and an even stronger leader!

GLOSSARY

activist - a person who emphasizes direct action in support of or in opposition to an issue that causes disagreement.

camouflage - when an animal's coloring matches its surroundings. Camouflage helps animals hide from predators and prey.

justice - the upholding of what is fair.

predator - an animal that lives by killing and eating other animals.

prey - an animal hunted or killed by other animals for food.

royalty - a person of royal blood and or status.

senses - a person's perception of sight, smell, hearing, taste, and touch.

stealth - done or acting in a secretive or covert way as not to be seen or heard.

ONLINE RESOURCES

Booklinks
NONFICTION NETWORK
FREE! ONLINE NONFICTION RESOURCES

To learn more about black panthers, please visit **abdobooklinks.com** or scan this QR code. These links are routinely monitored and updated to provide the most current information available.

INDEX